Chemical Engineering & the Law Forum 2019

Held at the 2019 AIChE Spring Meeting and
15th Global Congress on Process Safety

New Orleans, Louisiana, USA
31 March - 4 April 2019

ISBN: 978-1-5108-8699-5

Printed from e-media with permission by:

Curran Associates, Inc.
57 Morehouse Lane
Red Hook, NY 12571

proceedings.com

Some format issues inherent in the e-media version may also appear in this print version.

Printed by Curran Associates, Inc. (2019)

For permission requests, please contact AIChE
at the address below.

AIChE
120 Wall Street, FL 23
New York, NY 10005-4020

Phone: (800) 242-4363
Fax: (203) 775-5177

www.aiche.org

Additional copies of this publication are available from:

Curran Associates, Inc.
57 Morehouse Lane
Red Hook, NY 12571 USA
Phone: 845-758-0400
Fax: 845-758-2633
Email: curran@proceedings.com
Web: www.proceedings.com

TABLE OF CONTENTS

555706 Best Practices for Using Intellectual Property to Protect Chemical Process Inventions

To protect your chemical process innovations from competitors, it is vital to have a comprehensive plan to obtain and defend intellectual property ("IP"). A combination of patents, trademarks, and trade secrets can safeguard nearly any type of invention or information as long as you take the proper steps. This presentation will discuss what kinds of IP can be used to protect different kinds of inventions in the chemical industry. Speakers will also highlight several case studies addressing potential pitfalls and how to avoid them.

549176 Considerations and Best Practices in Licensing IP

This presentation will discuss basic considerations and best practices when licensing intellectual property. We will go over case studies and real world examples of issues that arise in licensing arrangements and how to resolve those issues.

550571 Aqualung Inferred: Inequitable Conduct in Regeneron Pharmaceuticals Inc. v Merus N.V

Inventors, attorneys and others involved in the filing and prosecution of a patent application owe a duty to disclose to the US Patent & Trademark Office any material information that could affect the patentability of the claimed subject matter. A granted patent can be held unenforceable for inequitable conduct where material prior art is withheld with an intent to deceive the USPTO.

Where inventors failed to disclose four material prior art references, an adverse inference of bad intent was drawn based on litigation misconduct.

AUTHOR INDEX